Pennie Stoyles and Christine Mulvany

The A–Z of Scientific Discoveries

Volume 3 H–L

Smart Apple Media

This edition first published in 2010 in the United States of America by Smart Apple Media.
All rights reserved. No part of this book may be reproduced in any form or by any means
without written permission from the publisher.

Smart Apple Media
P.O. Box 3263
Mankato, MN, 56002

First published in 2009 by
MACMILLAN EDUCATION AUSTRALIA PTY LTD
15–19 Claremont Street, South Yarra, Australia 3141

Visit our web site at www.macmillan.com.au or go directly to www.macmillanlibrary.com.au

Associated companies and representatives throughout the world.

Copyright © Pennie Stoyles and Christine Mulvany 2009

Library of Congress Cataloging-in-Publication Data

Stoyles, Pennie.
   The A to Z of scientific discoveries / Pennie Stoyles and Christine Mulvany.
      p. cm.
   Includes index.
   ISBN 978-1-59920-445-1 (hardcover)
   ISBN 978-1-59920-446-8 (hardcover)
   ISBN 978-1-59920-447-5 (hardcover)
   ISBN 978-1-59920-448-2 (hardcover)
   ISBN 978-1-59920-449-9 (hardcover)
   ISBN 978-1-59920-450-5 (hardcover)
   1. Discoveries in science--Encyclopedias, Juvenile. I. Mulvany, Christine. II. Title.
   Q180.55.D57S76 2010
   503--dc22
                     2009003443

Edited by Kath Kovac
Text and cover design by Ivan Finnegan, iF Design
Page layout by Ivan Finnegan, iF Design
Photo research by Legend Images
Illustrations by Alan Laver, Shelly Communications
Solar system illustrations (p.14) by Melissa Webb

Printed in the United States

Acknowledgments
The author and the publisher are grateful to the following for permission to reproduce copyright material:

Front cover photograph: Torch light © Asist/Dreamstime.com

Photos courtesy of: Digital Vision, **9**; © Adambooth/Dreamstime.com, **26** (right); © Asist/Dreamstime.com, **20** (right); © Naturevision_info/Dreamstime.com, **13** (top); © Neverhood/Dreamstime.com, **17**; Jonathan Alpeyrie/Getty Images, **25** (top); Monty Fresco/Topical Press Agency/Hulton Archive/Getty Images, **31**; Keystone/Getty Images, **8**; David Levenson/Getty Images, **13** (bottom); Hank Morgan - Rainbow/Getty Images, **20** (left); Purestock/Getty Images, **12**; James Sharples/Getty Images, **22** (bottom); Mansell/Time Life Pictures/Getty Images, **10**; © Andrey Volodin/iStockphoto, **21** (top); Library of Congress, **27**; Library of Congress © Oren Jack Turner, Princeton N.J., **28**; Photodisc, **29**; Photodisc/StockTrek, **4**, **14** (inset Jupiter); Photolibrary © martin norris/Alamy, **24**; Photolibrary/Brand X Pictures, **22** (top); Photolibrary/Science Photo Library, **25** (bottom); Photolibrary/Philippe Plailly/Eurelios/Science Photo Library, **7**; Photolibrary/Detlev Van Ravenswaay/Science Photo Library, **19**; Photolibrary/Superstock, **26** (left); Photos.com, **5** (right), **15** (bottom inset); © @erics/Shutterstock, **5** (left); © Stephen Girimont/Shutterstock, **15** (main behind); © Ronnie Howard/Shutterstock, **6**; Wikimedia Commons, photo by Glenlarson, **16**

While every care has been taken to trace and acknowledge copyright, the publisher tenders their apologies for any accidental infringement where copyright has proved untraceable. Where the attempt has been unsuccessful, the publisher welcomes information that would redress the situation.

Scientific Discoveries

Welcome to the Exciting World of Scientific Discoveries.

The A–Z of Scientific Discoveries is about the discovery and explanation of natural things. A scientific discovery can mean:
- finding or identifying something that exists in nature
- developing a theory that helps describe and explain a natural thing or event

A scientific discovery is sometimes the work of one person. Sometimes it is a series of discoveries made by many people building upon each other's ideas.

Volume 3 H–L Scientific Discoveries

Helium	4
Homo Sapiens	6
Hydrogen	8
Inheritance	10
Insulin	12
Jupiter	14
K-T Boundary	16
Kuiper Belt	18
Lasers	20
Laughing Gas	22
Leprosy (Hansen's Disease)	24
Lift	26
Light	28
Longitude	30

They Said It!

"The process of scientific discovery is, in effect, a continual flight from wonder."
Albert Einstein

Hh Helium

Helium is a colorless, odorless gas that is lighter than air. It is the second most common **element** in the universe, but is very rare on Earth.

How Helium Was Discovered

Helium was discovered on the sun before it was discovered on Earth. In 1868, French astronomer Pierre Janssen and British astronomer Norman Lockyer used the newly invented **spectroscope** to analyze the light coming from the sun. They saw that the sun was mostly made of hydrogen, but also noticed a second gas. Lockyer called this gas helium.

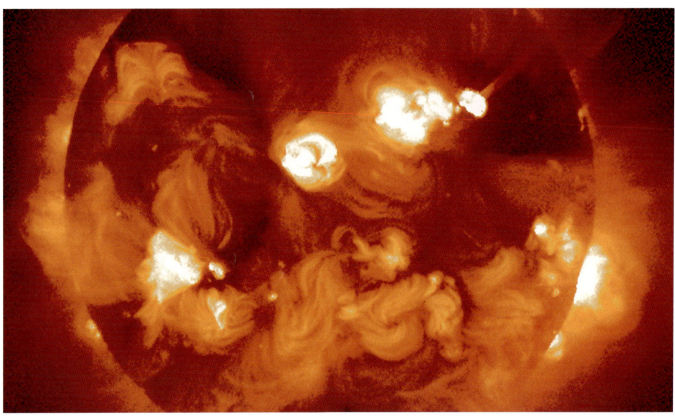

Helium gets its name from the Greek word for sun, helios.

Did You Know?

Helium is called a noble, or inert, gas because it does not react with anything else. Helium atoms never combine with other atoms to form new compounds.

Helium On Earth

In the 1890s, scientists knew that air was made mostly of nitrogen and oxygen, plus small amounts of other gases. In 1893, Scottish scientist Sir William Ramsay analyzed these gases with a spectroscope. He was very surprised to find that helium was found on Earth as well as on the sun.

Helium is lighter than air. Its chemical symbol is He, and it is element 2 because it has two protons in its nucleus.

Helium Balloons

Helium-filled balloons float because helium is lighter than air. Helium particles are so small that they can escape through the tiny pores in rubber balloons. This is why air-filled rubber balloons stay inflated much longer than those filled with helium.

Sir William Ramsay (1852–1916)

In addition to helium, Sir William Ramsay discovered four other gases in the air: argon, krypton, neon, and xenon. He won the Nobel Prize for Chemistry in 1904.

GLOSSARY WORDS

element   substance made of only one type of atom
spectroscope   an optical instrument used to measure properties of light

Homo Sapiens

Homo sapiens is the scientific name given to the animals known as modern humans.

Discovery of the Earliest *Homo Sapiens*

The oldest known skeletons of *Homo sapiens* date back 195,000 years. Kenyan **paleoanthropologist** Richard Leakey found the bones in Ethiopia in 1967. The bones were named Omo 1, because they were found near the Omo River.

Flake tools made of stone were found with the Omo 1 bones, showing that humans used tools 195,000 years ago. The first cultural artefacts, such as musical instruments and ornaments, found with early *Homo sapiens* skeletons date back only 50,000 years.

Our Nearest Living Relatives

Genetic studies have shown that related genes in chimpanzees and humans are 95 to 98 percent similar. However, there are many genes that the two species do not share. These are the genes that make us human and make chimpanzees apes.

Chimpanzees are the closest living relative of *Homo sapiens*.

Did You Know?
In 1619, Italian philosopher Lucilio Vanini was burnt at the stake for expressing his idea that modern humans evolved from apes.

Homo Sapiens and their Ancestors

By studying human evolution and the differences and similarities to other species, scientists can understand what is special about being human.

Homo sapiens are commonly thought to have evolved from ape-like ancestors. Modern humans and chimpanzees broke away from their common ancestor about six million years ago. Animals in the same **genus** as humans, such as *Homo habilis* and *Homo erectus*, evolved about two to three million years ago. *Homo sapiens* also lived alongside other human-like animals, such as *Homo neanderthalensis* (Neanderthal man), that later became extinct.

Homo habilis is thought to be the first animal with enough similarities to modern humans to be grouped in the same genus.

GLOSSARY WORDS

paleoanthropologist — a scientist who studies ancient humans
genus — group of species with similar characteristics that cannot interbreed

Hydrogen

Hydrogen is a colorless, odorless gas that is lighter than air. It is the most common **element** in the universe.

How Hydrogen Was Discovered

In 1766, English chemist Henry Cavendish made a new gas by dropping metals into different acids. He discovered that the gas was very light and that it exploded when lit, so he called it "flammable air." In 1788, French chemist Antoine Lavoisier found that when the gas exploded, it combined with oxygen to make water. He named the gas hydrogen, which means "water producer."

Flying With Hydrogen

In the 1700s, people were experimenting with hot-air balloons. French scientist Jacques-Alexander-Cesar Charles realized that because hydrogen is much lighter than air, it would work better in a balloon. Charles flew his hydrogen balloon for the first time in 1783. In the 1800s and early 1900s, people made airships by adding motors to hydrogen balloons.

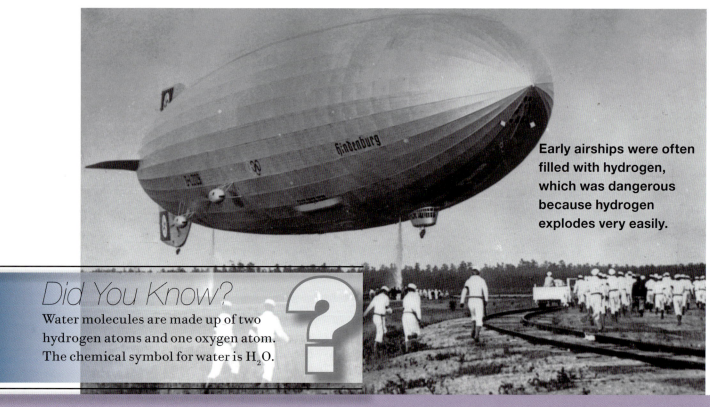

Early airships were often filled with hydrogen, which was dangerous because hydrogen explodes very easily.

Did You Know?
Water molecules are made up of two hydrogen atoms and one oxygen atom. The chemical symbol for water is H_2O.

Hydrogen in Everyday Life

Hydrogen is found in more substances than any other element on Earth. Most of Earth's hydrogen is found in water. Pure hydrogen can be extracted from water and used to make products such as fertilizers, margarine, and acids. Fuels such as oil, gasoline, and natural gas also contain hydrogen. The chemical symbol for hydrogen is H, and it is element 1 because it has one **proton** in its nucleus.

Future Fuel

Fuel cells are batteries that use hydrogen and oxygen to make electricity. They are being tested for use in electric cars and should soon be available for sale.

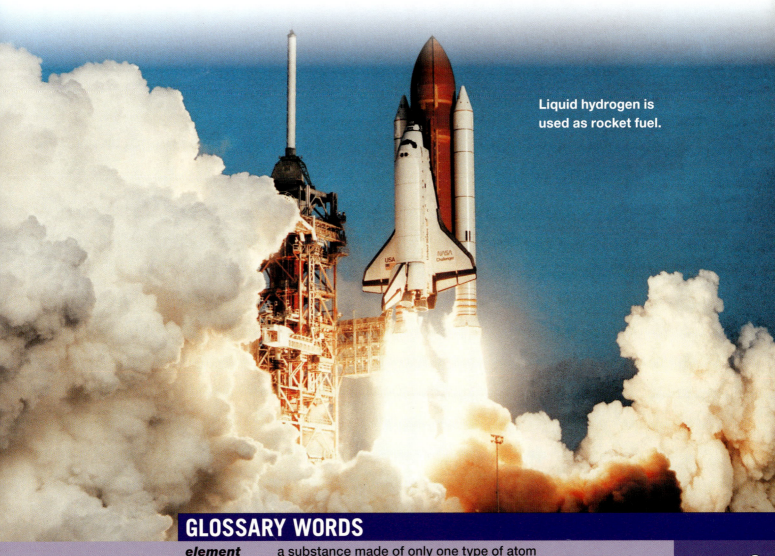

Liquid hydrogen is used as rocket fuel.

GLOSSARY WORDS
element   a substance made of only one type of atom
proton   particle found in the nucleus of all atoms

Ii Inheritance

Inheritance is the passing on of **characteristics** from one generation to another.

How Inheritance Was Discovered

Scientists started to study inheritance in the 1800s. They wanted to understand why different species of plants and animals exist, and why members of any one species are not exactly the same.

Between 1854 and 1868, Austrian monk Gregor Mendel did many breeding experiments using pea plants of different heights and different flower and seed colors. When he bred short pea plants together, he found that all the **offspring** were also short. When he bred tall plants together, or tall and short plants together, the offspring were either all tall or were a combination of tall and short. He could never produce a medium-sized plant. His careful analysis of the results explained a lot about how inheritance works.

Gregor Mendel (1822–1884) discovered how inheritance works. He is often called the "father of genetics."

How Inheritance Works

Mendel believed that each characteristic was associated with a "heredity factor." Today, we call these factors genes, and know that they are located on our **chromosomes**.

Mendel discovered that offspring inherit two heredity factors for each characteristic, one from each parent. He noticed that one of the heredity factors was sometimes "stronger" than the other. This is called dominant inheritance, and occurred in the case with the tall and short pea plants. The "tall" heredity factor, or gene, is dominant over the "short" gene.

For some characteristics, such as flower color, the offspring show a blend, or mixture, of characteristics from either parent. This is called intermediate inheritance.

Pea plant height is an example of dominant inheritance, and flower color is an example of intermediate inheritance.

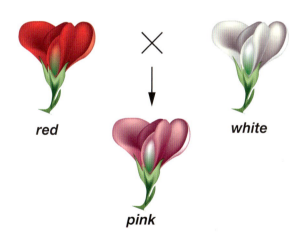

The offspring of red- and white-flowered pea plants will all have pink flowers (a blend of red and white).

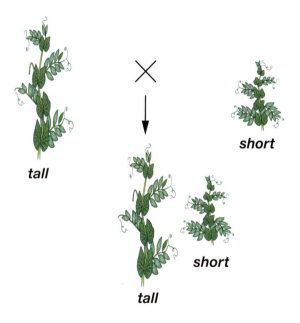

The offspring of tall and short plants will either be tall or short, never medium height.

Did You Know?
Mendel's theories were not immediately accepted. When he died in 1884, they were forgotten, and his work was not rediscovered until 1900.

GLOSSARY WORDS
characteristics features such as hair color or height
offspring the next generation; children are the offspring of their parents
chromosomes long molecules of DNA that hold genetic information

Insulin

Insulin is a **hormone** that controls the levels of sugar in our blood. People with diabetes cannot produce enough insulin and often need extra insulin to survive.

How Insulin Was Discovered

People knew about diabetes long before insulin was discovered. In the 1600s, **diabetics** were known to have sweet-tasting urine, and people knew that diabetes could be caused by damage to the pancreas.

In 1905, English **physiologists** Ernest Starling and William Bayliss discovered that the body made chemicals called hormones. Hormones act as messengers. They are released from one part of the body and travel to another part, where they control chemical processes in our cells. In 1916, another English physiologist, Edward Sharpey-Schafer, thought that the pancreas made a hormone that controlled the amount of sugar in our blood. In 1921, two Canadian researchers, Frederick Banting and Charles Best, finally discovered the insulin hormone. In the 1950s, British scientist Frederick Sanger discovered insulin's chemical structure.

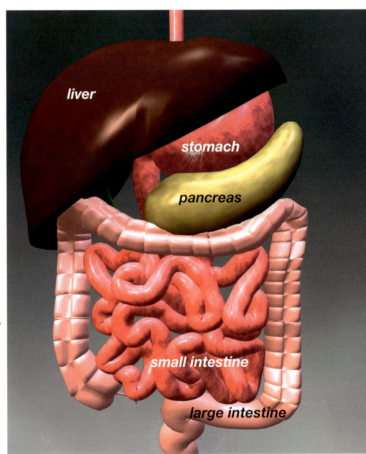

Insulin is made in the pancreas (yellow), which is located underneath the stomach (pink) and liver (brown), above the large intestine (light pink).

Did You Know?

The word insulin comes from the Latin word *insula*, for island. Insulin is made by special cells in the pancreas called islet cells.

Diabetes, Sugar, and Insulin

Your blood carries dissolved sugar to all the cells in your body, providing energy for them to work and grow. To stay healthy, your cells must take up a certain amount of sugar from the blood.

Insulin controls the take-up of sugar. As diabetics can not make enough insulin, their sugar levels are difficult to control. Sometimes, they can control their sugar levels through diet and exercise. Other diabetics must inject insulin into their bloodstream once or twice a day, every day of their lives.

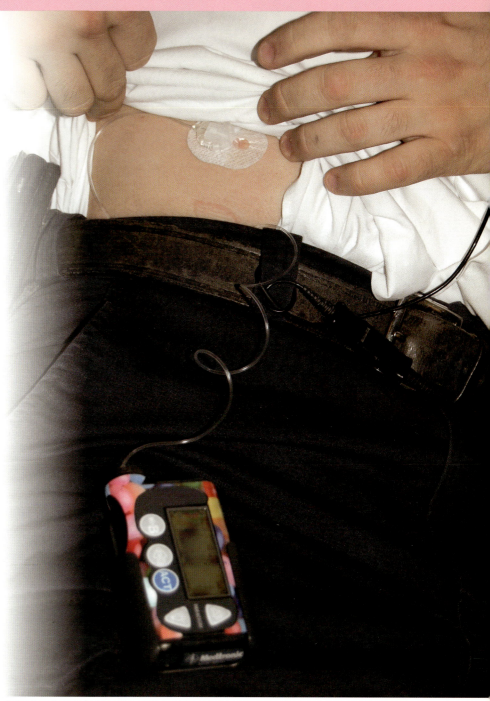

Diabetics sometimes wear computer-controlled insulin pumps to make sure their bodies receive the right amount of insulin.

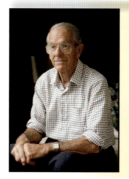

Frederick Sanger (born 1918)

Frederick Sanger is the only person to have been awarded two Nobel Prizes in Chemistry. He won his first in 1958 for discovering the chemical structure of insulin, and co-won his second in 1980 for working out how to read the information held in our genes.

GLOSSARY WORDS

hormone	a chemical messenger
diabetics	people with diabetes
physiologists	scientists who study how the body functions

Jupiter

Jupiter is the biggest planet in our solar system. It is the fifth planet from the sun.

The Discovery of Jupiter

Jupiter is usually the third-brightest object in the night sky, and can be seen without the use of a telescope. People have observed Jupiter since ancient times, so its discovery is not credited to just one person.

Jupiter is two-and-a-half times the mass of all the other planets in the solar system put together.

Jupiter Timeline

362–364 BC	1610	1973	1995
The Chinese astronomer Gan De is thought to have observed one of Jupiter's moons with the unaided eye	Using a telescope, Galileo Galilei discovered the four largest moons of Jupiter	The spacecraft *Pioneer 10* went within 80,788 miles (130,000 km) of Jupiter, collecting information about the planet's atmosphere	The spacecraft *Galileo* discovered that Jupiter's moon Europa may have an ocean beneath its surface capable of supporting life

New Discoveries About Jupiter

Telescopes and interplanetary spacecraft have led to more discoveries about Jupiter, which takes its name from the king of the ancient Roman gods. Four hundred years ago, people thought Jupiter had four moons. We now know it has at least 63 moons.

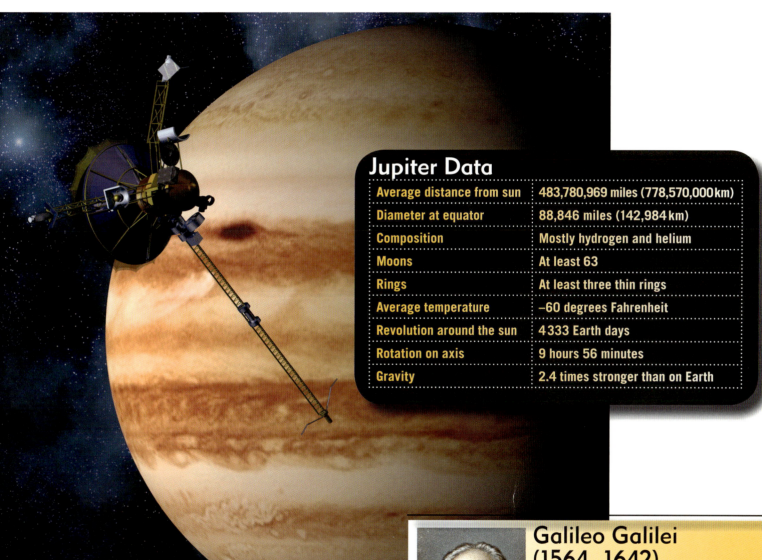

Jupiter Data

Average distance from sun	483,780,969 miles (778,570,000 km)
Diameter at equator	88,846 miles (142,984 km)
Composition	Mostly hydrogen and helium
Moons	At least 63
Rings	At least three thin rings
Average temperature	−60 degrees Fahrenheit
Revolution around the sun	4333 Earth days
Rotation on axis	9 hours 56 minutes
Gravity	2.4 times stronger than on Earth

In 1995, the spacecraft *Galileo Orbiter* released a probe into Jupiter's atmosphere that sent data back to scientists on Earth.

Galileo Galilei (1564–1642)

Galileo was an Italian scientist who improved the telescope and discovered the four largest moons of Jupiter: Io, Europa, Ganymede, and Callisto. His discovery opposed the popular belief of the time that all heavenly bodies revolved around Earth.

K–T Boundary

The K–T boundary is a thin layer of rock that formed 65 million years ago. It is found in many different parts of the world.

K–T Boundary and Mass Extinction

The K-T boundary is named after the geological times between which it formed. K is the abbreviation for the German word, *Kreidezeit*, for the Cretaceous Period (145-65 million years ago). T stands for the Tertiary Period (65-2 million years ago).

Scientists believe that the K-T boundary, where the K and T layers join, marks a time of mass **extinction**. This is because more than half of the once-living things (including dinosaurs) that are found as **fossils** in the Cretaceous (K) layer are not found in the higher Tertiary (T) layer.

By understanding how mass extinctions such as the one marked by the K–T boundary (red line) occurred, scientists think we can increase our chances of surviving another catastrophic event.

Did You Know?

An earlier extinction event happened, in which about 90 percent of species became extinct about 250 million years ago. It was known as the Permian–Triassic extinction event.

Discovering Clues to Extinction in the K–T Boundary

In 1980, scientists led by American physicist Luis Alvarez discovered that the K–T boundary contained unusually high levels of **iridium**. This metal is rare on Earth, but common in comets and asteroids. Alvarez predicted that a giant asteroid hit Earth 65 million years ago, and that the resulting dust from the impact settled to form the K–T boundary. In 1990, a huge impact crater called Chicxulub (say chik-shu-LUB) was found under the Yucatán Peninsula in Mexico. This confirmed Alvarez's theory.

The K–T boundary was probably created by dust from the impact of a giant asteroid, which is thought to be one of the factors that lead to the extinction of dinosaurs.

Effects of Asteroid Impact

Scientists think that the giant asteroid threw up massive amounts of dust and debris into the atmosphere when it hit Earth. This blocked the sunlight, which affected the growth of plant life and lowered the temperature, leading to mass extinctions. The impact may also have triggered volcanic eruptions, fires, tsunamis, and massive storms.

GLOSSARY WORDS
extinction — becoming extinct - no longer living
fossils — preserved remains or imprints of once-living things
iridium — a hard, brittle metal

The Kuiper Belt

The Kuiper (say KY-per) belt is a ring of icy objects left behind near the edge of the solar system after the planets formed.

The Discovery of the Kuiper Belt

The Kuiper belt's existence was predicted in the 1930s, but it has only recently been discovered. The belt's discovery required the development of equipment that could compare images of the same area of space, taken at different times. It also required a systematic search of the area near Pluto. In 1992, at Hawaii's Mauna Kea observatory, astronomers David Jewitt and Jane Luu discovered the first objects in the Kuiper belt, other than Pluto and its moon Charon.

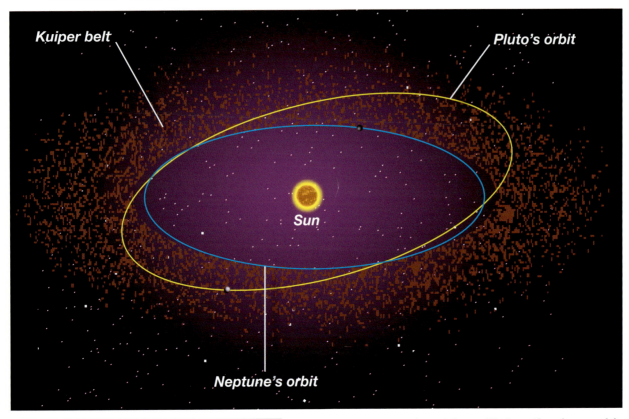

Did You Know?
At least nine other solar systems have rings of debris around them.

Apart from Pluto, other large objects in the Kuiper belt with a diameter over 62 miles (100 km) include: Orcus, Sedna, Charon, Quaoar, 2003 EL61, and Ixion.

More discoveries about the Kuiper belt are expected when the *New Horizons* spacecraft, launched in 2006, reaches Pluto in 2015.

More About the Kuiper Belt

The belt is named after Dutch-American astronomer Gerard Kuiper, one of several scientists who thought that small objects existed beyond Neptune.

The Kuiper belt is 30–50 **astronomical units** from the sun and contains at least 70,000 icy objects greater than 62 miles (100 km) across. These include Plutinos – icy bodies that orbit the sun at about the same rate as the dwarf planet Pluto, which is the largest known object in the Kuiper belt. The Kuiper belt is also thought to be the source of comets that take less than 200 years to complete an orbit of the sun. **Scattered disc objects** and **centaurs** are also thought to start from this area.

GLOSSARY WORDS

astronomical units	the average distance between Earth and the sun; about 93,206,000 miles (150 million km)
scattered disc objects	objects located further away than the Kuiper belt, with an unusual orbit
centaurs	icy bodies that orbit the sun between Jupiter and Neptune

Lasers

The term "laser" is an abbreviation of **l**ight **a**mplification by **s**timulated **e**mission of **r**adiation. Lasers produce a long, narrow beam of light that is made up of only one color.

How Lasers Were Discovered

In 1917, Albert Einstein thought of a way that lasers could work using a process called "stimulated **emission**." In 1947, American physicist Willis Lamb discovered how stimulated emission worked, but he did not make a laser. Scientists then made a kind of laser, called a maser, which used microwaves instead of light. In 1960, another American physicist Theodore Maiman used the maser as a starting point to make the first working laser.

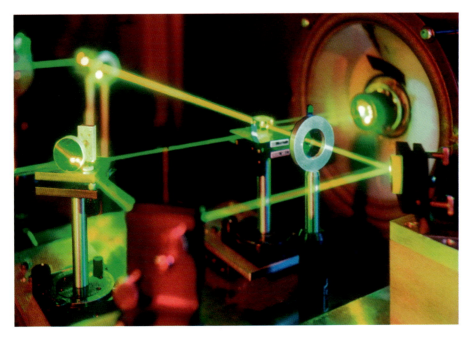

Laser beams, such as this one in a research laboratory, are long and do not spread out.

Theodore Maiman (1927–2007)

When Theodore Maiman invented the laser, he did not know what it would be useful for, saying "A laser is a solution seeking a problem." He was nominated for a Nobel Prize twice for his work on lasers, but did not receive one.

Normal light beams, such as from a flashlight, are short and spread out quickly.

Lasers in Everyday Life

Lasers have many different uses. They are used in laser printers, and to scan product barcodes. Lasers cut microscopic pits on the surface of CDs and DVDs, and also read the discs in CD and DVD players.

Lasers can have different power levels. Barcode readers, disk players, and laser pointers use low-power lasers. Doctors use more powerful lasers in some types of surgery to make fine cuts, remove skin blemishes, and improve eyesight. The most powerful lasers are used to cut plastics and metals.

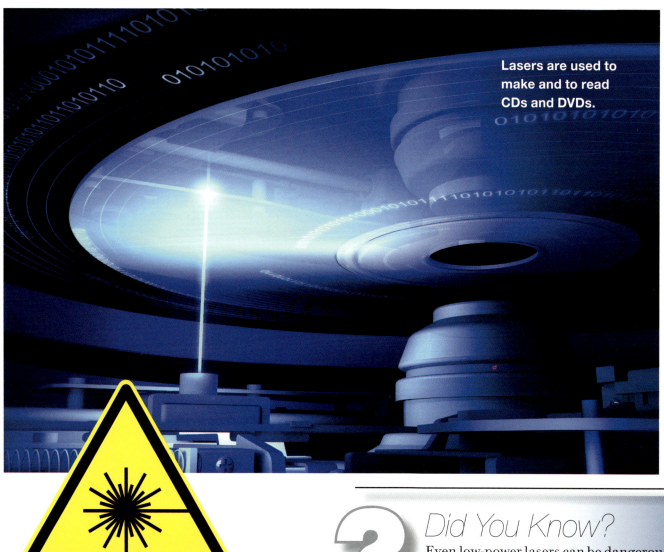

Lasers are used to make and to read CDs and DVDs.

The laser hazard symbol warns people when high-power lasers are in use.

Did You Know?
Even low-power lasers can be dangerous to human eyesight. Never look directly at a laser beam or point a laser at anyone's face.

GLOSSARY WORD

emission   sending out

Laughing Gas

Laughing gas is also called nitrous oxide. It is a colorless gas with a slightly sweet odor and taste. Dentists often use it as an **anesthetic**.

How Laughing Gas Was Discovered

Nitrous oxide was discovered in 1772 by English scientist Joseph Priestley. In 1799, another English scientist, Sir Humphrey Davy, named it laughing gas when he saw that people who breathed in nitrous oxide laughed, danced around, and acted silly. Davy also realized that laughing gas was an anesthetic, but his discovery was ignored for more than 40 years.

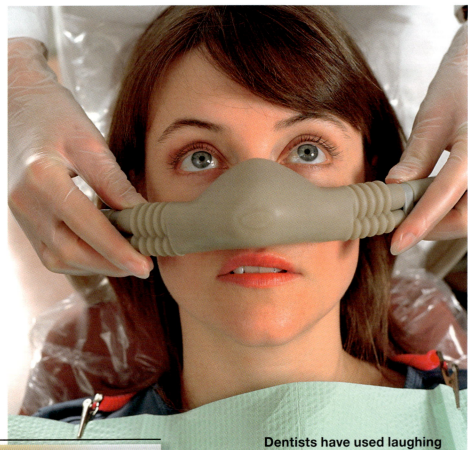

Dentists have used laughing gas as an anesthetic for more than 150 years, and they still use it today to dull pain and calm anxious patients.

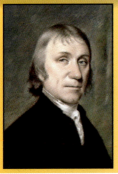

Joseph Priestley (1733–1804)

Priestley was a Presbyterian minister who made many discoveries about electricity and chemistry. He discovered oxygen and invented soda water. He is sometimes called the "father of chemistry."

An Accidental Discovery

When laughing gas was first discovered, it was often used as entertainment at parties and in carnival sideshows. People would pay to breathe in a small amount of the gas, or to watch others doing it.

In 1844, American dentist Dr. Horace Wells watched a man breathe in laughing gas at an exhibition. The man staggered back and sat next to Dr. Wells, who saw he had a bad cut. When the dentist asked the man how he received the cut, he said he did not know, because he could feel no pain. Dr. Wells realized that laughing gas dulled pain, and demonstrated its use in dentistry by having one of his own teeth removed while **inhaling** the gas.

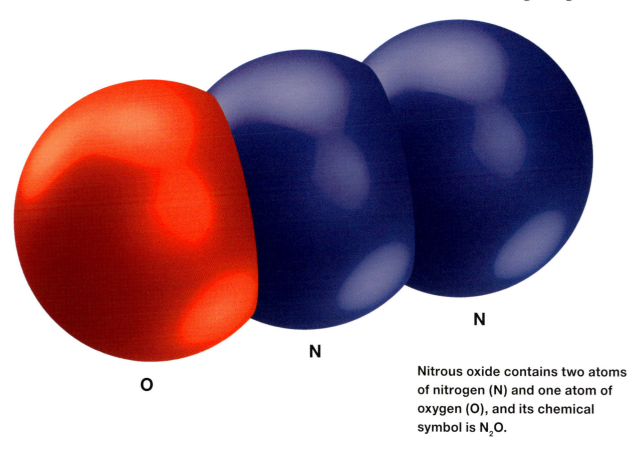

Nitrous oxide contains two atoms of nitrogen (N) and one atom of oxygen (O), and its chemical symbol is N_2O.

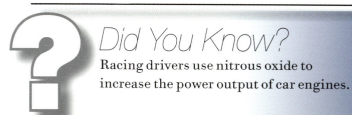

Did You Know?
Racing drivers use nitrous oxide to increase the power output of car engines.

GLOSSARY WORDS

anesthetic   substance that dulls pain
inhaling   breathing in

Leprosy (Hansen's Disease)

Leprosy affects the skin and nerves, particularly on the face, feet, and hands. Untreated leprosy patients can become deformed, crippled, and blind.

Leprosy in Ancient Times

Humans have endured leprosy since ancient times. The earliest written records date back to 1350 BC. In the past, it was known as "living death." People with leprosy were taken from their communities and placed in isolated leper colonies.

Discovering the Cause and Cure for Leprosy

In 1870, Norwegian doctor Gerhard Hansen identified brown elements in the cells of leprosy patients as the cause of the disease. Ten years later, German doctor Albert Neisser identified the brown elements as a type of bacteria called *Mycobacterium leprae*. In the 1950s, a medicine called dapsone was used to kill the bacteria, but the bacteria soon became **resistant** to it. In the 1970s, Indian scientist Shantaram Yawalkar discovered a combination of medicines that provided a cure for leprosy.

People with leprosy used to be forced to live in isolated leper colonies, but today, this once-feared disease is completely curable.

Did You Know?
In 2007, about 4.5 million people were affected by leprosy. Most of the cases were found in Brazil, India, Madagascar, Mozambique, and Nepal.

Catching Leprosy

Leprosy is spread when people with the disease cough or sneeze, sending the bacteria into the air. Of every 100 people who become infected with the disease, about 95 will be able to fight off the bacteria themselves. The remaining people will require treatment.

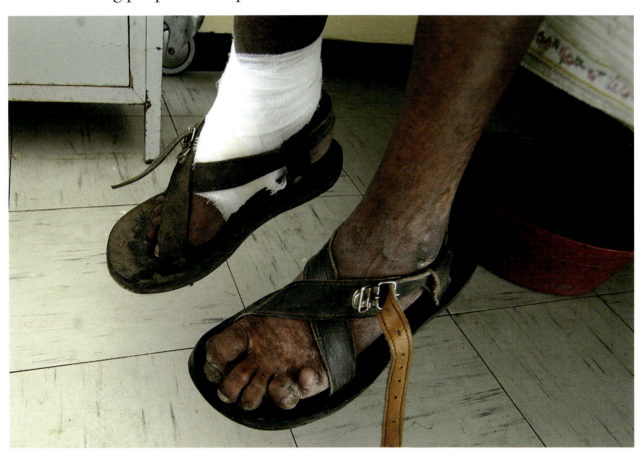

Leprosy patients lose feeling and do not react to dangerous situations, so they can easily cut or burn themselves.

A World Without Leprosy

In 1985, a research team at Stanford University began developing a **vaccine** against leprosy. In the near future, scientists hope to prevent the disease from occurring, rather than just treating it.

Dr. Gerhard Henrik Armauer Hansen (1841–1912)

Hansen spent his life studying the cause of leprosy and caring for patients. Once the cause of leprosy was identified, it was renamed Hansen's disease in his honor.

GLOSSARY WORDS

resistant   unable to be killed
vaccine   a preparation that helps the body fight disease

Lift

Lift is the force generated when a solid object moves through liquid or gas. The force of lift is at a right angle to the movement.

Discovering the Principles of Lift

The amount of lift created depends on the object's speed measured as if the gas or liquid were stationary. The size and shape of the object affects the lift, as does the **inclination** at which it is moving. Understanding lift requires an understanding of the laws of motion, as well as knowledge of how liquids and gases interact. Sir Isaac Newton played an important role when he developed the three principles of motion in 1686. Dutch scientist Daniel Bernoulli also helped to develop our understanding of liquids and gases.

The amount of lift depends on the airspeed and the angle, shape, and material of the kite.

If the wind is flowing horizontally, then the force of lift will move the kite vertically.

Daniel Bernoulli (1700–1782)

Daniel Bernoulli was a great mathematician, medical doctor, and scientist. He developed mathematical equations to explain his observations about lift.

Discovering Equations to Calculate Lift

In 1901, Orville and Wilbur Wright realized that the lift equations of the day were inaccurate. The kites, gliders, and planes they built based on these equations did not create the lift expected. They built a **wind tunnel** to test more than 50 models and came up with their own lift equation, which they used to build the first successful powered aircraft.

Orville and Wilbur Wright discovered the lift equation that allowed them to build the first powered aircraft.

Lift in Everyday Life

Objects that move through air or water rely on the force of lift. This includes the wings on planes and the rotors on helicopters.

GLOSSARY WORDS	
inclination	tilt; angle between two planes
wind tunnel	an area where air is blown over objects at known speeds

Light

Light is a form of energy. Light waves travel in tiny particles called photons.

How Light Was Discovered

In the 1600s, scientists developed two different theories about the behavior of light. Sir Isaac Newton believed that light was a stream of particles, but other scientists thought that it traveled in waves. These theories were debated for hundreds of years until 1905, when Albert Einstein suggested that light could behave like both waves and particles. He said that light travels in "bundles of waves," which are now called photons.

In 1921 Albert Einstein received the Nobel Prize in Physics for his work on the behavior of light.

Light Timeline

1021	1666	1905	1923
Arabic mathematician Ibn al-Haytham discovered that light from the sun reflects from objects into our eyes	Sir Isaac Newton discovered that white light was a combination of colored light	Albert Einstein developed his "bundle of waves" or quantum theory of light	Arthur Crompton used the word photon to describe bundles of light waves

The Speed of Light

In 1676, Danish astronomer Ole Roemer timed one of Jupiter's moons as it **eclipsed** Jupiter. He noticed that when Jupiter was closer to Earth, the eclipse was earlier than predicted, and when Jupiter was further away, the eclipse was later than predicted. He used his observations to work out the speed of light as 141,051 miles (227,000 km) per second. This was quite close to today's figure of 186,282 miles (299,792 km) per second.

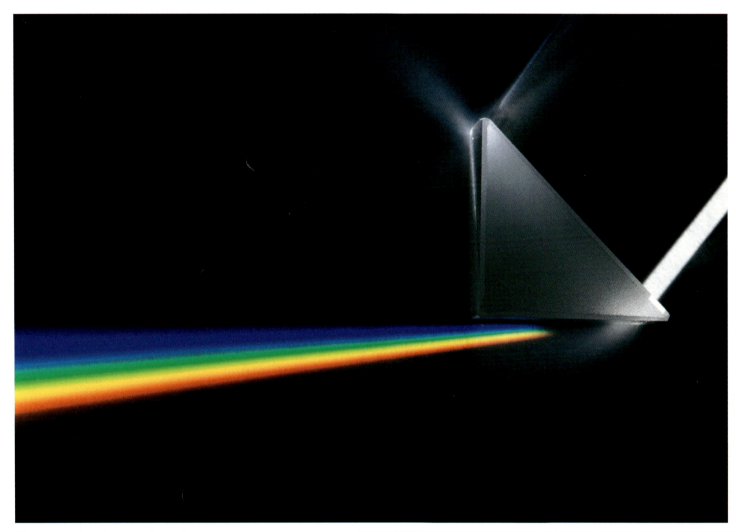

Sir Isaac Newton used a prism to discover that white light was a mixture of other colors.

Did You Know?
In 1838, German astronomer Frederick Bessel first used the term "light year" to describe the distance that light travels in one year. It is around 5.9 **trillion** miles (9.5 trillion km).

GLOSSARY WORD

eclipsed — passed in front of
trillion — a million million million

Longitude

Longitude is the distance east or west of an imaginary line that connects the North and South Poles. It is used in **cartography** and **navigation**.

Lines of Longitude

Longitude is measured in degrees. The lines of longitude range from zero degrees at a line called the prime meridian to 180 degrees to the east (+) or west (-) of that line. Each degree (°) is divided into 60 minutes (') and each minute is divided into 60 seconds ("). The line that passes through Greenwich, England, was chosen as the international prime meridian, or zero line of longitude, in 1884.

Why Measuring Longitude Is Important

Until an accurate way of determining longitude was discovered in the 1700s, sailors navigated by adding the distance and direction of each day's travel. This was often inaccurate over long voyages and many ships took the long route, staying near coastlines. On long voyages, when not in sight of land, some became lost or were shipwrecked.

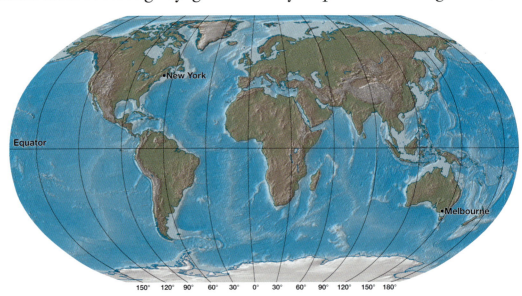

Melbourne's longitude east of the prime meridian is +144° 58', and New York's longitude to the west is −73° 58'.

Did You Know?
Today's sailors navigate using a system that calculates the ship's position from time signals transmitted by satellites.

How Measuring Longitude Was Discovered

In 1553, mathematician and cartographer Gemma Frisius suggested that a clock could measure longitude. Because the Earth rotates 360 degrees in 24 hours, one hour represents 15 degrees of longitude. Sailors could therefore determine longitude by comparing their local time with the time at the zero longitude. Although the idea was sound, it wasn't until 1736 that English clockmaker John Harrison developed a clock called a marine chronometer that was reliable at sea.

Using the Skies to Find Longitude

In 1514, German mathematician Johannes Werner proposed using the moon's position to work out longitude. By 1766, astronomers had produced a table of calculations known as the Nautical Almanac. The almanac gave sailors a way of working out the time at zero longitude after making measurements of the Moon and sun or other stars. This lunar distance was only popular until 1850.

Marine chronometers, used for navigation in the 1800s, had to keep accurate time in different temperatures, at different angles, and in the presence of moisture and salt.

GLOSSARY WORDS
cartography — map-making
navigation — finding your way

Index

Page references in bold indicate that there is a full entry for that discovery.

A
airships 8
Alvarez, Luis 17
asteroid impact 17

B
bacteria 24
Banting, Frederick 12
Bayliss, William 12
Bernoulli, Daniel 26
Best, Charles 12

C
Cavendish, Henry 8
Charles, Jacques-Alexander-Cesar 8
Chicxulub crater 17
chromosomes 11
chronometer 31
Cretaceous Period 16

D
Davy, Sir Humphrey 22
diabetes 12, 13

E
Einstein, Albert 20, 28

F
Frisius, Gemma 31
fuel cells 9

G
Galilei, Galileo 14, 15
Galileo 14
Gan De 14

H
Hansen, Gerhard Henrik Armauer 24, 25
Harrison, John 31
helium **4-5**
Homo sapiens **6-7**
humans 6, 7
hydrogen **8-9**

I
inheritance **10-11**
insulin **12-13**

J
Janssen, Pierre 4
Jewitt, David 18
Jupiter **14-15**, 29

K
K–T boundary **16-17**
Kuiper belt **18-19**
Kuiper, Gerard 19

L
Lamb, Willis 20
laser **20-21**
laughing gas **22-23**
Lavoisier, Antoine 8
Leakey, Richard 6
leprosy **24-25**
lift **26-27**
light **28-29**
light, speed of 29
Lockyer, Norman 4
longitude **30-31**
Luu, Jane 18

M
Maiman, Theodore 20
mass extinction 16, 17
Mendel, Gregor 10, 11

N
Neisser, Albert 24
Newton, Sir Isaac 26, 28, 29
nitrous oxide 22, 23
noble gases 4, 5

O
Omo 16

P
pancreas 12
Pluto 18, 19
Priestley, Joseph 22
prime meridian 30

R
Ramsay, Sir William 5
Roemer, Ole 29

S
Sanger, Frederick 12, 13
Sharpey-Schafer, Edward 12
Starling, Ernest 12

T
Tertiary Period 16

V
vaccine 25

W
Wells, Horace 23
Werner, Johannes 31
Wright, Orville and Wilbur 27

Y
Yawalkar, Shantaram 24